Visit Susie and her Animal Friends at
KidsKyngdom.com

For my 3 Children

CAMILET

That's a mix up of
Cameron, Dominic & Violet

Thank You SO Much for
all of Your Support!

A Very Special Thank You to **Ceola**

She Reviewed this Animal Safari Book and Gave Fantastic 5-Year-Old Feedback!

Visit Susie and her Animal Friends at
KidsKyngdom.com

Thank you for supporting
Kids Kyngdom

Get Your **FREE**
audio book
at
KidsKyngdom.com

Hi
I'm Susie

Doctor Susie

if you please

Do you have a Nickname?

Sometimes I go by Dr. Smash

Today we're going on an **Animal Safari**

to learn all about **frogs**

Visit Susie and her Animal Friends at
KidsKyngdom.com

Bullfrogs

to be even more Specific!

The frogs I know live in a river

They lay their eggs
In the water

A bunch of frog eggs are called

a **frogspawn**

Not **Song**

Visit Susie and her Animal Friends at
KidsKyngdom.com

When the spawn hatch they turn into baby tadpoles

I call them **Squirmers**

Because they have No Lungs

Just Gills Like Fish

They make me

Happy

Soon they **grow up** into

brown, green and sometimes yellow **Big** frogs

What should we call this one?

How about Sebastian

Bash for short

If you have a **nickname**

what would it be?

Hey Bash
Did you know

you can't drink **any** water?

Visit Susie and her Animal Friends at
KidsKyngdom.com

Oh no Bash

RIBBIT

You can't **Sip** it

Visit Susie and her Animal Friends at
KidsKyngdom.com

That's right
You need to **soak** it in

Through your **Skin**

If you dry out

You **just** might

That's Right

Visit Susie and her Animal Friends at
KidsKyngdom.com

Frog eyes rotate
All around

This helps them
catch their **prey**

Not Hay

Visit Susie and her Animal Friends at
KidsKyngdom.com

Prey

Prey is their food

They love **Bugs**

Visit Susie and her Animal Friends at
KidsKyngdom.com

They catch **bugs**

with Really **Long** tongues

Come on **Bash**

What animal should we **visit** next?

RiBBiT

Visit Susie and her Animal Friends at
KidsKyngdom.com

A Rabbit?

How about

an even Bigger Animal?

Like a **Lion**

Lions have Amazing
hearing

Do you hear that?
A Lion is calling us
from **Africa**

Come on **Bash**
Let's go **visit** Lions

Did you Love
visiting Frogs
with Susie and Bash?

Please Share Your Review

Meet Sammie Kyng

Sammie is a mother of 3 and a lover of nature and animals

She is the proud creator of Dr. Susie, a young scientist - *Sammie's alter-ego* - and her new friend Bash (He's a **Frog**)

Here she is at her home in in Minnesota, USA

Join Susie and Bash as they study animals, critters, bugs, furry (and slimy) things and more at:

KidsKyngdom.Com   Instagram.com/Kids_Kyngdom

ISBN: 9978-1-959501-00-8

Published by: Kyngdom, LLC

Do you know who the
Best Hunter is
in a Pride of Lions?

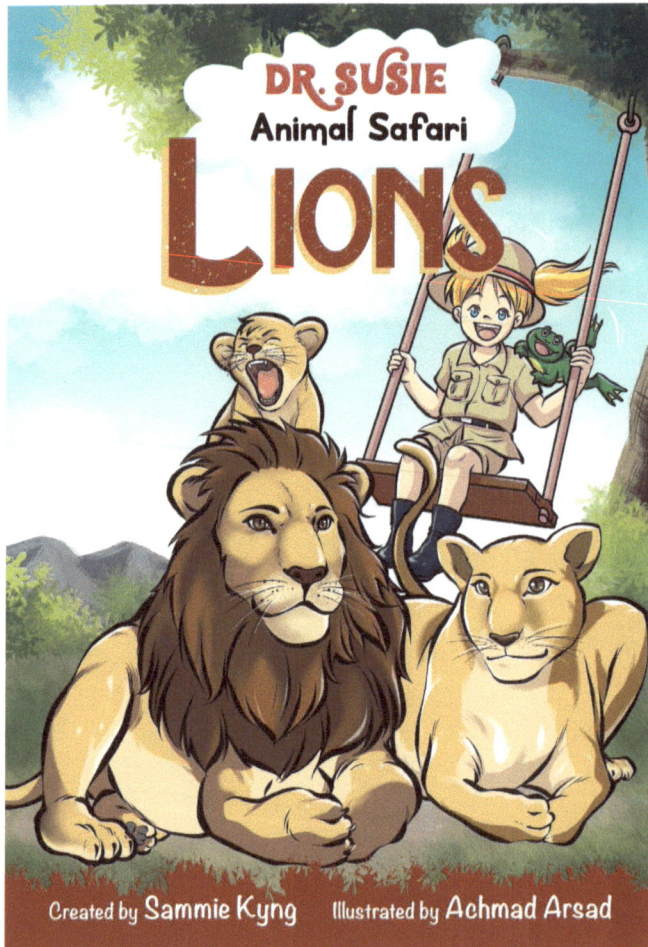

Let's learn about it with
Dr. Susie and Bash!